ENERGY SECTOR STANDARD OF THE PEOPLE'S REPUBLIC OF CHINA

中华人民共和国能源行业标准

Code for Construction Organization Design of Onshore Wind Power Projects

陆上风电场工程施工组织设计规范

NB/T 31113-2017

Chief Development Department: China Renewable Energy Engineering Institute
Approval Department: National Energy Administration of the People's Republic of China
Implementation Date: March 1, 2018

China Water & Power Press
中国水利水电出版社
Beijing 2024

All rights reserved. No part of this publication may be reproduced, stored in a retrieval system, or transmitted in any form or by any means—electronic, mechanical, photocopying, recording or otherwise, without prior written permission of the publisher.

图书在版编目（CIP）数据

陆上风电场工程施工组织设计规范：NB/T 31113-2017 = Code for Construction Organization Design of Onshore Wind Power Projects(NB/T 31113-2017)：英文 / 国家能源局发布. -- 北京：中国水利水电出版社, 2024. 10. -- ISBN 978-7-5226-2773-1

Ⅰ. TM614-65

中国国家版本馆CIP数据核字第20249UF822号

ENERGY SECTOR STANDARD
OF THE PEOPLE'S REPUBLIC OF CHINA
中华人民共和国能源行业标准

Code for Construction Organization Design
of Onshore Wind Power Projects
陆上风电场工程施工组织设计规范
NB/T 31113-2017
（英文版）

Issued by National Energy Administration of the People's Republic of China
国家能源局　发布
Translation organized by China Renewable Energy Engineering Institute
水电水利规划设计总院　组织翻译
Published by China Water & Power Press
中国水利水电出版社　出版发行
　　Tel: (+ 86 10) 68545888　68545874
　　sales@mwr.gov.cn
　　Account name: China Water & Power Press
　　Address: No.1, Yuyuantan Nanlu, Haidian District, Beijing 100038, China
　　http: //www.waterpub.com.cn
中国水利水电出版社微机排版中心　排版
北京中献拓方科技发展有限公司　印刷
184mm×260mm　16开本　2印张　63千字
2024年10月第1版　2024年10月第1次印刷
Price（定价）：￥320.00

Introduction

This English version is one of China's energy sector standard series in English. Its translation was organized by China Renewable Energy Engineering Institute authorized by National Energy Administration of the People's Republic of China in compliance with relevant procedures and stipulations. This English version was issued by National Energy Administration of the People's Republic of China in Announcement [2023] No. 5, dated October 11, 2023.

This version was translated from the Chinese Standard NB/T 31113-2017, *Code for Construction Organization Design of Onshore Wind Power Projects*, published by China Water & Power Press. The copyright is reserved by National Energy Administration of the People's Republic of China. In the event of any discrepancy in the implementation, the Chinese version shall prevail.

Many thanks go to the staff from the relevant standard development organizations and those who have provided generous assistance in the translation and review process.

For further improvement of the English version, any comments and suggestions are welcome and should be addressed to:

China Renewable Energy Engineering Institute
No. 2 Beixiaojie, Liupukang, Xicheng District, Beijing 100120, China
Website: www.creei.cn

Translating organization:

POWERCHINA Northwest Engineering Corporation Limited

Translating staff:

FU Zhiqiang	TIAN Weihui	CUI Zhenlei	WANG Dandi
ZHANG Ge	ZHAO Yue	XUE Aiguo	TIAN Yongjin
ZHOU Kang	NAN Ya		

Review panel members:

LIANG Hongli	Shanghai Investigation, Design & Research Institute Co., Ltd.
QI Wen	POWERCHINA Beijing Engineering Corporation Limited
WANG Lei	POWERCHINA Huadong Engineering Corporation Limited

LI Yu	POWERCHINA Huadong Engineering Corporation Limited
LI Kejia	POWERCHINA Northwest Engineering Corporation Limited
JIA Haibo	POWERCHINA Kunming Engineering Corporation Limited
YOU Yang	China Society for Hydropower Engineering
Hu Yongzhu	POWERCHINA Northwest Engineering Corporation Limited
Wu Hupo	China Resources Power Holdings Co.,Ltd.

National Energy Administration of the People's Republic of China

翻译出版说明

本译本为国家能源局委托水电水利规划设计总院按照有关程序和规定，统一组织翻译的能源行业标准英文版系列译本之一。2023年10月11日，国家能源局以2023年第5号公告予以公布。

本译本是根据中国电力出版社出版的《陆上风电场工程施工组织设计规范》NB/T 31113—2017 翻译的，著作权归国家能源局所有。在使用过程中，如出现异议，以中文版为准。

本译本在翻译和审核过程中，本标准编制单位及编制组有关成员给予了积极协助。

为不断提高本译本的质量，欢迎使用者提出意见和建议，并反馈给水电水利规划设计总院。

地址：北京市西城区六铺炕北小街2号
邮编：100120
网址：www.creei.cn

本译本翻译单位：中国电建集团西北勘测设计研究院有限公司

本译本翻译人员：付志强　田伟辉　崔振磊　王丹迪
　　　　　　　　张　戈　赵　悦　薛爱国　田永进
　　　　　　　　周　康　南　雅

本译本审核人员：

梁洪丽　上海勘测设计研究院有限公司

齐　文　中国电建集团北京勘测设计研究院有限公司

王　蕾　中国电建集团华东勘测设计研究院有限公司

李　瑜　中国电建集团华东勘测设计研究院有限公司

李可佳　中国电建集团西北勘测设计研究院有限公司

贾海波　中国电建集团昆明勘测设计研究院有限公司

由　洋　中国水力发电工程学会

胡永柱　中国电建集团西北勘测设计研究院有限公司

吴琥珀　华润电力控股有限公司

国家能源局

Announcement of National Energy Administration of the People's Republic of China [2017] No. 10

According to the requirements of Document GNJKJ [2009] No. 52, "Notice on Releasing the Energy Sector Standardization Administration Regulations (*tentative*) and detailed implementation rules issued by National Energy Administration of the People's Republic of China", 204 sector standards such as *Safety Management Specification for Coalbed Methane Production Stations*, including 62 energy standards (NB), 86 electric power standards (DL), and 56 petroleum standards (SY), are issued by National Energy Administration of the People's Republic of China after due review and approval.

Attachment: Directory of Sector Standards

National Energy Administration of the People's Republic of China

November 15, 2017

Attachment:

Directory of Sector Standards

Serial number	Standard No.	Title	Replaced standard No.	Adopted international standard No.	Approval date	Implementation date
...						
10	NB/T 31113-2017	Code for Construction Organization Design of Onshore Wind Power Projects			2017-11-15	2018-03-01
...						

Foreword

According to the requirements of Document GNKJ [2009] No. 163 issued by National Energy Administration of the People's Republic of China, "Notice on Releasing the Development and Revision Plan of the First Batch of Energy Sector Standards in 2009", and after extensive investigation and research, summarization of practical experience, and wide solicitation of opinions, the drafting group has prepared this code.

The main technical contents of this code include: construction transportation, engineering construction, general construction layout, and master construction schedule.

National Energy Administration of the People's Republic of China is in charge of the administration of this code. China Renewable Energy Engineering Institute has proposed this code and is responsible for its routine management. Sub-committee on Construction and Installation of Wind Power Project of Energy Sector Standardization Technical Committee on Wind Power is responsible for the explanation of specific technical contents. Comments and suggestions in the implementation of this code should be addressed to:

China Renewable Energy Engineering Institute
No. 2 Beixiaojie, Liupukang, Xicheng District, Beijing 100120, China

Chief development organization:

POWERCHINA Northwest Engineering Corporation Limited

Participating development organization:

Hebei Electric Power Design & Research Institute

Chief drafting staff:

HU Yongzhu	YANG Jing'an	CUI Zhenlei	LI Jiandang
SHEN Kuanyu	GUO Zongqiang	ZHANG Peng	GUAN Qinghua
WANG Liping	QIN Chusheng	MENG Jinbo	MA Jianchun
SONG Yan			

Review panel members:

| YI Yuechun | CHEN Huiming | CHANG Zuowei | TANG Huan |
| ZHANG Quan | WU Chaoyue | LI Qinwei | CHEN Guibin |

XIE Yanli	GUO Shijie	GE Xiaobo	ZHOU E'na
WANG Mingtao	YAN Xi	WANG Jingli	ZENG Jie
LI Yujie	CONG Ou	LI Chao	LI Shisheng

Contents

1	**General Provisions**	1
2	**Basic Requirements**	2
3	**Construction Transportation**	3
3.1	Transportation Options	3
3.2	Site Access	3
3.3	On-Site Roads	3
4	**Engineering Construction**	5
4.1	Civil Works	5
4.2	Equipment Installation	6
4.3	Collection Line Construction	7
5	**General Construction Layout**	8
5.1	Auxiliary Facilities	8
5.2	Site Layout	8
5.3	Cut and Fill Balance and Disposal Area Planning	9
5.4	Construction Land	9
6	**Master Construction Schedule**	11
6.1	Master Construction Schedule Plan	11
6.2	Resources Allocation	12
Appendix A	**Main Basic Data Required in Construction Planning**	13
Appendix B	**Major Technical Indexes for Construction Transportation**	14
Explanation of Wording in This Code		16
List of Quoted Standards		17

1　General Provisions

1.0.1　This code is formulated with a view to standardizing the construction organization design of onshore wind power projects, to improve the design level and ensure the design quality.

1.0.2　This code is applicable to the construction organization design for the construction, renovation, and extension of onshore wind power projects.

1.0.3　In addition to this code, the construction organization design of onshore wind power projects shall comply with other current relevant standards of China.

2 Basic Requirements

2.0.1 The construction planning of onshore wind power projects shall follow the principles below:

1. Harmonize the construction with surrounding environment by taking the local conditions into consideration.

2. Achieve rational economic and technical indicators by following specified construction procedures, proper construction sequence, overall planning, and effective construction tempo.

3. Adopt new technologies, materials, processes, and equipment.

4. Advocate environment-friendly, safe and civilized construction methods.

5. Realize efficient land use by considering the local land use planning.

2.0.2 The design considerations of the construction planning shall include:

1. Requirements of the state and local government related to the project construction.

2. Overall planning of the wind farm.

3. Project owner's requirements.

4. Construction conditions, traffic conditions, and recent development plans of the region where the project is located.

5. Technical conditions and related equipment supplies.

6. Design documents.

2.0.3 The construction planning shall meet the requirements of environmental protection and soil and water conservation.

2.0.4 The main basic data required in construction planning shall comply with Appendix A of this code.

3 Construction Transportation

3.1 Transportation Options

3.1.1 The transportation scheme and route shall be determined according to the project location, equipment transportation requirements, transportation conditions, and connection of access roads and on-site roads.

3.1.2 The transportation scheme for large and heavy pieces shall consider the following factors:

 1 Size and weight of large and heavy pieces such as the nacelle, tower, blades, hub, and main transformer for transportation.

 2 Capacities of hauling equipment.

 3 Road capacity.

3.1.3 A road construction, renovation or extension scheme or a temporary bypass scheme shall be put forward according to the selected transportation route.

3.2 Site Access

3.2.1 The site access shall meet the requirements of materials and equipment transportation and operation management.

3.2.2 The site access shall be designed by considering the factors such as the topographical and geological conditions of the project area, local transportation requirements, and transportation intensity during construction.

3.2.3 The main technical indexes of the site access should be in accordance with Article B.0.1 of this code.

3.3 On-Site Roads

3.3.1 The layout of on-site roads shall consider the following factors:

 1 Topography and geology, arrangement of civil structures, construction layout, social environment, etc.

 2 Features of large and heavy pieces and transportation requirements.

 3 Capacities of hauling equipment.

 4 Status of on-site roads.

 5 Operation and maintenance requirements.

3.3.2 Necessary protection facilities and traffic signs shall be set for on-site roads.

3.3.3 The construction road used for upcoming maintenance shall meet relevant requirements.

3.3.4 The main technical indexes of on-site roads should be in accordance with Article B.0.2 of this code.

4 Engineering Construction

4.1 Civil Works

4.1.1 The selection of construction scheme shall consider the following factors:

1 Hydrometeorology, topography and geology.

2 Type of foundation and structures.

3 Material supply.

4 Type selection and arrangement of construction machinery.

5 Surrounding social environment.

4.1.2 The earth and rock excavation shall meet the following requirements:

1 The foundation excavation shall be performed in layers from top to bottom.

2 The impact crushing method should be adopted for the rock excavation of foundation.

4.1.3 The earth and rock backfilling shall meet the following requirements:

1 The excavated materials shall be given priority for backfilling.

2 The backfill of foundation shall be compacted in layers.

4.1.4 For the pile foundation construction, appropriate construction machinery shall be selected according to the pile type.

4.1.5 The concrete raw materials shall be selected according to material supply, concrete performance requirements, construction conditions in the project area and other factors.

4.1.6 The concrete mix proportion shall satisfy the design properties and construction requirements.

4.1.7 Truck mixers should be used for the transportation of concrete. The construction procedures such as concrete mixing, transportation, placement, and temperature control shall be connected reasonably.

4.1.8 The concrete for wind turbine foundation shall be placed continuously without construction joints.

4.1.9 For the foundation concrete placement, temperature control measures should be taken and the construction requirements in rainy and cold seasons shall be met.

4.1.10 The concrete placed shall be cured in time by watering, covering, or spraying curing agent.

4.1.11 For the wind turbine foundations and towers, if new materials or processes are used, the construction scheme shall be demonstrated specifically.

4.2 Equipment Installation

4.2.1 The selection of installation scheme shall consider the following factors:

1 Hydrometeorology, topography and geology.

2 Dimensions, weight and installation location of components and parts.

3 Type selection and arrangement of installation machinery.

4.2.2 The stockpiling of wind turbines shall meet the following requirements:

1 The stockyard shall be leveled and well drained, and meet the requirements for bearing capacity and lightning protection.

2 The main parts should be stockpiled in the order of installation and placed within the working range of crane.

4.2.3 The wind turbine installation conditions shall meet the following requirements:

1 Roads shall be leveled and unobstructed to facilitate safe passage of various construction vehicles.

2 A sufficient space for storage and assembly shall be provided.

3 The age of foundation concrete shall not be less than 28 d or the strength shall not be less than 75 % of the design value.

4 The installation of foundation earthing network is completed.

4.2.4 The main installation equipment shall be selected according to the installation schedule and unloading and installation requirements of wind turbines, taking into account the outline dimensions, center of gravity, piece weight, and installation height of the main parts of wind turbines.

4.2.5 The installation schedule for wind turbines shall coordinate with the foundation construction. The installation of wind turbines shall meet the requirements of manufacturers and avoid thunderstorm.

4.2.6 The installation of main electrical equipment shall meet the following requirements:

1 The installation schedule shall coordinate with the construction

sequence of civil works to avoid interference.

2   The embedded parts should be buried along with the construction of structural concrete.

3   The installation shall not commence until the foundation concrete reaches 70 % of the design strength.

4.2.7   Appropriate construction methods for cable laying and lightning-protection earthing shall be selected according to the laying type and construction sequence.

4.3   Collection Line Construction

4.3.1   The construction layout shall be determined according to the collection line design, taking into account the route, traffic, crossing, spanning, stockyard, and water and power supply conditions, etc.

4.3.2   The transportation method shall be determined according to the local construction conditions. The road width should be no less than 1.2 m and slope no steeper than 1 : 4 for hand haulage.

4.3.3   For towers located nearby traffic crossings, slopes or riversides, the warning signs and protective measures shall be provided according to the site-specific conditions .

4.3.4   The line tensioning shall be conducted only after the tower foundation concrete has reached the design strength and the towers within the entire tensioning range have been accepted.

4.3.5   Buried cables in the frozen soil region should be laid below the frozen soil layer, if not applicable, the cables may be laid in the dry frozen soil layer with good permeability or backfill soil, or other measures to prevent the cables from damage may be adopted as well. When crossing railways or roads, the cables shall be laid in solid protective pipes.

5 General Construction Layout

5.1 Auxiliary Facilities

5.1.1 The supply scheme of aggregates and concrete shall be determined according to local conditions.

5.1.2 Sites of aggregate processing plant and concrete batching plant shall be determined according to the traffic conditions, water and power supply conditions, concrete construction requirements, etc., and the production capacity of the plants shall satisfy the demand of concrete placement in construction peak month.

5.1.3 The water supply scheme shall be determined according to the water source conditions, and the supply capability and quality shall meet the requirements of production and living water consumption.

5.1.4 The construction should adopt the power supply from the local power grid, and shall satisfy the peak demands of production and living power consumption.

5.1.5 The communication system for construction should use the local communication network.

5.2 Site Layout

5.2.1 The construction site should be zoned by functions, and may be divided into the construction area and the living area. These two areas should be separated from each other.

5.2.2 The construction site shall avoid socially or environmentally sensitive areas, cultural relics and unfavorable geological sites, and should be centralized, make use of wastelands and sloping lands, and minimize the use of cultivated lands.

5.2.3 The use of construction site should be coordinated with the construction sequence and the site should be reused.

5.2.4 Layout of the erection platform shall meet the requirements of stockpiling, assembling and installing the wind turbine components and parts.

5.2.5 The floor area of construction management and living facilities is calculated by multiplying the average headcount in peak month by per capita floor area that may take 6 m^2 to 8 m^2.

5.2.6 Flood control measures shall be taken for both construction and living areas, the flood recurrence interval should be between 5 and 20 years, and the main construction and living areas should take the upper limit.

5.3 Cut and Fill Balance and Disposal Area Planning

5.3.1 The disposal area planning shall be based on the result of cut and fill balance.

5.3.2 The cut and fill balance planning should use the excavated materials to reduce disposal. The utilization amount and disposal of excavated materials shall be determined by their properties. The backfills shall be stored separately from the disposals.

5.3.3 The disposal area arrangement shall facilitate the transportation with short haul distance for both backfills and disposals, and shall avoid geological hazards such as landslide, debris flow, karst, and water inrush.

5.3.4 The height and slope ratio of the disposal body shall meet the stability requirements. Diversion, drainage and retaining facilities shall be arranged for the disposal area according to the flood control requirements.

5.3.5 The flood control standard of disposal area shall be determined according to the scale of the disposal area and the degree of hazard after failure. The flood control standard shall comply with the current national standard GB 51018, *Code for Design of Soil and Water Conservation Engineering*.

5.4 Construction Land

5.4.1 The construction land use shall be determined in line with the scientific, rational and economical principles.

5.4.2 The land area of various buildings, facilities, and roads shall be determined with consideration of the local land policy, and the lands for permanent use and temporary use shall be identified.

5.4.3 The calculation of permanent land use area should meet the following requirements:

1. The land area of wind turbines is calculated based on the outer contour size of the wind turbine foundation.

2. The land area of box-type substations is calculated based on the outer contour size of the substation foundation.

3. The land area of cable trenches is the product of the total length of the cable trenches and 1.5 m.

4. The land area of overhead lines is calculated based on the outer contour size of the tower foundation.

5. The land area of the step-up substation is calculated to 1 m beyond the fence wall.

6 The land area of the access roads and the roads for maintenance during operation is calculated based on the subgrade width.

7 The rebuilt and existing roads are not included in the land area calculation, while for the expanded roads, the increased area may be considered in the calculation.

8 For other permanent works constructed for the safety and operation of the wind farm, the land area is calculated based on the actual situation.

5.4.4 The calculation of temporary land use area should meet the following requirements:

1 The land area of crane pad is calculated based on the actual area proposed in the construction layout.

2 The land area of the construction road is calculated based on the outer contour size of the horizontal projection of temporary structures for transportation.

3 The land area of the buried cables is the product of the total length of the cable trenches and 1 m.

4 For other facilities such as temporary office and accommodation facilities, workshops, stockyard of aggregates, concrete batching plant, machinery and equipment stockyard, repair shops, warehouses, wind turbines and tower equipment temporary stockyard, and disposal area, the land area takes the actual area proposed in the construction layout.

5.4.5 In the case that the construction roads are used for upcoming maintenance, or the land for wind turbine assembly and erection are also used for civil works such as wind turbine foundation, box-type substation foundation, and earthing, the land area shall be counted only once.

6 Master Construction Schedule

6.1 Master Construction Schedule Plan

6.1.1 The master construction schedule plan shall be prepared according to the project features, scale, technical complexity and construction capacity.

6.1.2 The construction period of an onshore wind power project shall be divided into the pre-preparatory period, preparatory period, and main works construction period. The preparatory period starts from the issue of Notice to Proceed and ends before the main works foundation construction begins, and the subsequent main works construction period ends on the Commercial Operation Date. The total construction period is the sum of the preparatory period and the main works construction period.

6.1.3 The preparation of master construction schedule plan shall meet the following requirements:

1 Comply with the fundamental construction procedures of the state.

2 Arrange the construction period according to the current average advanced construction capacity, and consider the impact of geology, climate, transportation, social environment, etc.

3 Determine the schedule plans for key items such as step-up substation construction, wind turbine foundation construction, and wind turbine installation based on the type selection of construction machinery and equipment.

4 Coordinate the individual item schedule with the master schedule, well arrange the construction procedures to facilitate construction and reduce interference.

5 Arrange the erection, debugging, start-up and test run of wind turbines according to the supply plan of wind turbines and main transformer.

6 Balance the resources allocation.

6.1.4 The schedule plans of preparatory period and main works construction period shall be coordinated with each other and the construction works may be reasonably cross-arranged.

6.1.5 The main works construction schedule shall be determined according to the construction equipment capacity, effective construction time, construction intensity, etc.

6.2 Resources Allocation

6.2.1 The construction labor index shall be determined according to the domestic average advanced construction level and the project-specific conditions.

6.2.2 The supply plan of main construction materials such as cement, steel and concrete aggregates shall be determined according to the master construction schedule.

6.2.3 The resources allocation shall be optimized according to the master construction schedule plan and the supply plan for the main construction machinery and equipment shall be proposed.

Appendix A   Main Basic Data Required in Construction Planning

Table A   Main basic data required in construction planning

No.	Description	Required data
1	Natural conditions	1 Hydrological and meteorological data measured in the project area 2 Topographical, geomorphological, and geological conditions of the project area
2	Social conditions	1 Local supply of construction materials and living goods 2 Land use and land requisition conditions of the construction area 3 Supporting facilities construction conditions in the project area
3	Project conditions	1 Requirements and approval opinions of the local government and related competent authorities for project construction 2 Previous design results and data 3 Current situation of site access and transportation of outsourced materials 4 Water source, water and power supply in the construction area
4	Material supply	1 Data on the sources and supply conditions of construction materials 2 Production test or field test results of new materials, processes, and technologies
5	Construction resources	1 Current situation of local construction markets and labor supply 2 Performance indicators and capacities of construction machinery and equipment

Appendix B Major Technical Indexes for Construction Transportation

B.0.1 The main technical indexes of access roads should be in accordance with Table B.0.1.

Table B.0.1 Main technical indexes of access roads

Item	Index
Design driving speed (km/h)	30
Minimum radius of horizontal curve (m)	40
Maximum longitudinal slope (%)	8
Maximum slope length (m)	300
Minimum radius of convex (concave) vertical curve (m)	400
Minimum length of vertical curve (m)	25
Pavement width (m)	6.5
Design flood frequency for subgrade	1/25

NOTES:
1. For special vehicles, the minimum radius of horizontal curve may take 0.6 to 0.8 times the design value.
2. The maximum longitudinal slope of individual road sections in poor conditions may be increased by 2 %.
3. When the maximum longitudinal slope is less than 6 %, there is no limit to the slope length.
4. The maximum slope length refers to the allowable length under the maximum longitudinal slope condition.

B.0.2 The main technical indexes of on-site roads shall be in accordance with Table B.0.2.

Table B.0.2 Main technical indexes of on-site roads

Item	Index
Design driving speed (km/h)	15
Minimum radius of horizontal curve (m)	40
Maximum longitudinal slope (%)	12
Maximum slope length (m)	300
Minimum radius of convex (concave) vertical curve (m)	400

Table B.0.2 *(continued)*

Item		Index
Minimum length of vertical curve (m)		20
Design flood frequency for subgrade		1/10
Pavement width (m)	One-lane	4.5
	Dual-lane	6.5

NOTES:

1 For special vehicles, the minimum radius of horizontal curve may take 0.6 to 0.8 times the design value.

2 The maximum longitudinal slope of individual road sections in poor conditions may be increased by 3 %.

3 When the design value of the maximum longitudinal slope exceeds the value in this table, the transportation safety should be demonstrated.

4 When the maximum longitudinal slope is less than 6 %, there is no limit to the slope length.

5 The maximum slope length refers to the allowable slope length under the maximum longitudinal slope condition.

Explanation of Wording in This Code

1   Words used for different degrees of strictness are explained as follows in order to mark the differences in executing the requirements in this code.

    1) Words denoting a very strict or mandatory requirement:

    "Must" is used for affirmation; "must not" for negation.

    2) Words denoting a strict requirement under normal conditions:

    "Shall" is used for affirmation; "shall not" for negation.

    3) Words denoting a permission of a slight choice or an indication of the most suitable choice when conditions permit:

    "Should" is used for affirmation; "should not" for negation.

    4) "May" is used to express the option available, sometimes with the conditional permit.

2   "Shall meet the requirements of…" or "shall comply with…" is used in this code to indicate that it is necessary to comply with the requirements stipulated in other relative standards and codes.

List of Quoted Standards

GB 51018,   *Code for Design of Soil and Water Conservation Engineering*